E-Learning for Educators

E-Learning for Educators

Denise Taylor

E-Learning for Educators

Publisher: Independent Publishing Network.Publication date: November 2020ISBN: **978-1-83853-918-4**Author: Denise TaylorDistributor: Pressbooks, https://pressbooks.com/Please direct all book orders or enquiries to the distributor.Copyright © 2020 Denise Taylor All rights reserved. No part of this book may be reproduced in any form or by any electronic or mechanical means, including information storage and retrieval systems, without permission in writing from the publisher, except by reviewers, who may quote brief passages in a review. ISBN 978-1-83853-918-4 Printed in United Kingdom

Contents

Main Body 1
Digital Technologies 1
Functionality within Technologies 1
Using APPs in Teaching and Learning 1
Using Video effectively in Teaching and Learning 1
Using Online Platforms in Teaching and Learning 1
Using Applications Software in Teaching and Learning 1
Conclusion 1
About the author A 21st Century E-Learning Guide for Teachers, Trainers and Instructors Denise Taylor 1
https://support.google.com/youtube#topic=9257498 1
https://www.khanacademy.org/computing/computer-programming/programming#animation-basics 1
Introduction to E-Learning 1

2.
3.
4.
5.
6.
7.
8.
9.
10.
11.

Introduction to E-Learning

INTRODUCTION

E-Learning

The 2020 pandemic has catapulted the style and practice of teaching and learning into the tech arena quite forcefully. Many educators and learners have been caught off-guard for various reasons. In short, teaching and learning now largely occurs using a variety of electronic technologies. Many teachers and learners have had to catch up and many find themselves at a loss on how to teach and learn to the best of their abilities.

E-Learning involves learning using electronic technologies, mostly using hardware, such as, a computer, tablet, phablet or mobile phone that is connected to the internet. In other words, a SMART device. These SMART devices connects teachers and learners to a large variety of sources of information, in formats, such as, text, audio, music, videos, chat-type facilities (social media), etc.

The ease of access to many outstandingly useful resources changes the traditional dynamics between teachers and learners. It could mean that teachers are no longer believed to be the 'fount of all knowledge' because expertise could come from a source (through the internet) other than the teacher. 'Experts' can now be invited into a lesson via VOIP (voice over internet protocol) technologies, such as, SKYPE, ZOOM, etc. to enhance the quality of teaching and learning.

Learners can upload their exercises onto an online classroom portal and learners can peer-review and improve their own exercises, thereby raising standards between learners as they learn from each other.

Learners can also work on joint projects together remotely online. This teaching and learning style has many benefits for learners as they work on sections of a project that matches their individual abilities more closely than previous learning expectations that all learners should achieve the same levels of understanding. Such teaching methods personalizes and differentiates learning for all ability levels of learners by allowing learners to do sections of a joint project where they are able to explore their individual curiosities best.

Technology is everywhere and whether we like it or not, it's here to stay.

This book will equip you with the computer skills to empower you and make you more confident to use 21st century teaching styles in your teaching and training programmes. You will learn about the functionality on your digital device, explore what's available on the internet for your subject, how to use and create video more effectively, how online teaching and learning platforms integrate technologies that can enhance collaborative learning and working styles and discover which APPs are most useful for your teaching needs.

Whether you're an accomplished educator who is probably used to juggling several balls in the air at once, or a trainee teacher, don't miss out on this opportunity to keep up with the *Tech Generation.*

This book will suit educators who are teachers, instructors or trainers in all sectors of education. It is aimed at equipping teachers who wish to embrace the use of computer technologies more fully to engage and enhance their teaching and learning practices.

The approach in this book, is different from simply reading it from start to finish. In order the gain the maximum benefit from this book, you should prepare to be an *active learner by doing the practice teaching exercises as they arise in each chapter.*

Requirements for Active Participation

- a reliable internet connection
- one or more electronic devices, such as, a computer, laptop, tablet, mobile/cell phone
- email account for signing up for online resources
- students to practise your lessons
- teaching colleagues for collaborative tasks and feedback

Topics you will cover in this book include:

- E-Learning for the 21st century teaching styles
- Using your digital device for teaching and learning

- Using and creating video for teaching and learning
- How can online teaching and learning platforms and VLEs be used to enhance collaborative learning and working styles
- Exploring APPs for teaching your subject
- Engaging uses for QR codes
- Discovering the effectiveness of using online surveys in teaching and learning
- Discovering how collaborative working styles raise standards between peers
- Catering for students remotely using online face-to-face technologies
- Enjoying the benefits of using free Open Educational Resources (OER)
- Accessing the benefits for using standard applications software in teaching and learning

Digital Technologies

1

In this chapter, you will:

- explore E-Learning styles that involve using digital technologies in teaching and learning
- evaluate typical 21st century lesson plan styles and approaches

Introduction

E-Learning is learning using an electronic device and involves connecting to the internet to access resources and communicate with others in the teaching and learning experience. There are pockets of excellent examples where e-learning is taking place to great effect and there are times when e-learning is not an effective way to teach or learn. This can be because of the Digital Divide. However, the 21st century has introduced amazing ways to use different technologies to enhance the teaching and learning experience all around the world.

Digital technologies can be described as both the hardware and software that teachers and learners use to access their teaching and learning content and resources. The hardware includes devices, such as, laptops, desktop computers, tablets, phablets or mobile SMART phones. The software includes applications, such as, word processors, spreadsheets, presentations, etc., online applications, such as, online surveys (Survey Monkey), virtual learning environments (VLEs), voice-over-internet-protocols (VOIP), such as SKYPE, ZOOM, Google Meet, etc., massive online open courses (MOOCs), such as, Khan Academy, EDX, Future Learn, etc., social media sites, such as, FaceBook, Twitter, Instagram, etc., online collaborative tools, such as, PREZZI, TRELLO and online tutorials, such as, www.w3schools.com and www.codecademy.com, etc. and Open Educational Resources (OER), such as, YouTube, Ted Talks, BBC Bitesize, etc. This list is not exhaustive and is constantly being revised, updated and added to.

Examples of Tasks using Digital Technologies

Scenario 1

Task 1 below is a typical example of the types of questions and tasks found in classrooms today, where references are made to online resources, such as, finding information by using search engines, such as, Google, Yahoo, etc. and using links to videos posted online at places, such as, YouTube, etc.

Task 1

- *Using Google, find 3 definitions for the term, 'e-learning'. Compose your own definition for the term, 'E-Learning'.*
- *List 5 different digital technologies that you could use in a teaching and learning context. Paste images found on the World Wide Web.*
- *Watch this video "Revolutionizing Education from the ground up" and comment on your whether you agree or disagree with Sir Ken Robinson's ideas on the state of education in today's world.*

Discussion

You could use question 1 as a **Starter** activity and use a **wiki** where learners have to edit the first effort at defining the term, 'É-Learning'. As each learner enters what they believe is their definitive answer, the previous answer is replaced until a final definition is adopted by everyone. Such an exercise encourages learners to think deeply about each word they choose to use in their definitions and to question whether their definition, encapsulated a comprehensive enough effort on their part. There are many benefits for engaging learners more actively, especially if a replacement definition doesn't appear good enough in their eyes. The list of benefits is not exhaustive, as you may experience many more benefits or perhaps some drawbacks as well.

Question 2 throws up some issues of copyright and has unexpected benefits or outcomes besides familiarising learners with the initial information about different digital technologies, such as, "Have I acknowledged the sources of my images? Have I got permission to use these images?"

Question 3 is good for setting questions that learners could answer, discuss or debate afterwards.

Scenario 2

In Task 2, you will find that many people simply go to www.YouTube.com when they want to find out how to do almost anything. Question 1, highlights an important aspect of how we learn anything, i.e. by watching someone else doing something and then trying to do that same thing and variations of it ourselves.

Question 2 expects you to model the use of the technologies in a teaching and learning context. You could use SKYPE in a live transmission to connect to an expert baker to demonstrate how to bake a cake. Learners stuck in a classroom or at home on their devices would absolutely be enthralled. I'm sure you could think of a myriad of ways to use some of the technologies in even more creative ways than mentioned here.

Task 2

- Consider using digital technologies to learn how to play a musical instrument, bake a cake or fix a leaking tap. Describe how the process of learning to do any of these things using digital technologies compares to learning them the traditional ways without computers?
- Select one of the learning items (how to play the musical instrument, bake the cake or fix the leaking tap) and create a spider diagram showing how you would use:
    - Open Educational Resources (OER), such as, YouTube.
    - Voice Over Internet Protocol (VOIP), such as, SKYPE, ZOOM, etc.
    - Social Media, such as, FaceBook, Pinterest, Twitter, Messenger, etc

Discussion

What does it mean to 'model' in a teaching and learning context? A teacher models a task by pretending to be a learner. Typical language during a modelling exercise is, "If I were you, I would …". In the example of using www.YouTube.com to learn how to bake a cake or transmitting a live SKYPE demonstration from a baker, students will not only learn how to bake a cake but also how to use www.YouTube.com and SKYPE to learn about how to do certain things.

Consider how you would use YouTube, SKYPE or ZOOM, WhatsApp or FaceBook, etc. to enhance the quality of teaching and learning in your lessons.

Scenario 3

It's important to refer to the latest academic research publications in order to keep up to date with the latest educational trends, much of which includes some aspect of technology. Similar articles such as those reflecting Bloom's Taxonomy and how the levels can be achieved using digital technologies are explained by Niall McNulty on November 27, 2017

https://www.niallmcnulty.com/2017/11/blooms-digital-taxonomy/

Task 3

Refer to this article on 'Digital Learning Literacies' in order to answer the next question

https://www.researchgate.net/publication/318508429_Digital_Literacy_for_the_21st_Century

Referring to the excerpt above taken from the article above MATCH these digital literacies with the appropriate digital technologies in Task 2 number 2:

Spires and Bartlett (2012) have divided the various intellectual processes associated with digital literacy into three categories:

- *locating and consuming digital content*
- *creating digital content*
- *communicating digital content*

Discussion

Categorising technologies according to how they could be used when preparing a lesson, can help you to choose the most effective technology for the desired outcome. If you want students to create digital content, the more appropriate technologies for this task could be using video, audio, image or text types of technologies. As creating, constructing, assembling or designing are very high level thinking skills, learners will have to pull together a multitude of skills, such as, finding information, organising the information and arranging it into some sequence to create their output.

Scenario 4

It is assumed that lesson plans basically consist of three to four main parts: a starter, a main teaching part, a time for learners to carry out their tasks and a plenary. There are many variations to a basic lesson plan. Integrated into these different part of a lesson plan can be a host of other phenomena, such as, assessment for learning or assessment of learning, metacognition tasks, differentiation, facilitation of the learning process, etc.

Task 4

Evaluate the lesson plan below in terms of its use of e-learning elements, such as, digital technologies used by the teacher and the digital literacies being developed by the students. How does it compare with traditional 'chalk-n-talk' teaching and learning styles?

Exemplar Lesson Plan

Lesson objectives:

To learn about the first 20 elements in the Periodic Table.

Starter

Play the 'Periodic Table Song' twice and ask students to complete this short quiz:

https://www.surveymonkey.co.uk…

Main Teaching

- Ask students to use their computer devices to research information on the Periodic Table and to create a presentation showing only the first 20 elements of the Periodic Table.

- Ask students to compile 10 True or False questions at the end of their presentation.
- Ask students to show their presentation to one of their friends and ask them to answer their 10 questions. Add their friend's scores and comments in the last slide of their presentation.

Plenary

Ask students

- List all of the first 20 elements of the periodic table in order.

- How well did I learn today?

- Is there another strategy that I could use to learn more effectively?

Discussion

The digital technologies being used in this example lesson plan involves the learners accessing the web link for the starter that the teacher created to check their learner's prior knowledge of this lesson's content. Learners will require a device to link onto the internet to do the survey first. The teacher will be able to see how well the students know or do not know the first 20 elements of the Periodic Table as soon as each student has completed their online quiz during the starter part of the lesson. This could be likened to a traditional starter where a teacher could have printed out the same questions and asked students to answer them during the starter exercise. However, if the class size is large, the teacher would not have had enough time to ascertain how well students knew or didn't know about the Periodic Table as quickly as doing it using an online tool, such as, Survey Monkey.

During the main teaching part of the lesson, the teacher takes a 'back seat' as learners engage more actively with finding out as much as they can from the many OERs available online. This frees up the teachers time to help less confident learners. The task that learners need to complete is to prepare 10 TRUE or FALSE questions and place them is a presentation (PREZZI). Assessment for learning is in-built as learners will be able to swap their presentations with their peers to try to answer each other's questions online.

The plenary will reveal how well learners were able to achieve the lesson outcomes. Learners and the teacher could discover how many elements they knew from the first 20 elements of the Periodic Table and discuss ways they might be able to learn it more effectively next time.

In low-tech schools, the same materials could be downloaded prior to a lesson and some materials could be printed out and shared among learners. There are various options for delivering the same lesson depending on which end of the Digital Divide the teaching and learning occurs.

Practical Task for teachers

- Using the same lesson plan style used in the example in Task 4, create a lesson plan for a topic in a subject of your choice. Use as many digital technologies as you can.
- Explain which digital literacies you are developing in your students for each type of digital technology that you have used in your lesson plan.
- Share your lesson plan with a colleague and ask for their feedback.

Functionality within Technologies

2

In this chapter, you will:

- explore the uses of the features of digital devices
- critically review and then create a teaching and learning event using the features of a digital device
- evaluate the usefulness of the device's features for teaching and learning

Introduction

The plethora of digital devices can sometimes make it difficult to know which one is the best for teaching and learning. Broadly, a computer device that allows teachers and learners to write text, record audio and visual content and one that connects to the internet is sufficient to enjoy most teaching and learning experiences. It could be a laptop, a desktop computer, a phablet, a tablet or a SMART mobile phone. Storage devices such as portable hard drives, USB pen drives or cloud storage should be sufficient to store your digital content.

The challenge amidst all the noise of so many digital devices and digital applications is to focus on the job of teaching so that your learners learn. It is worth thinking about how this quote might apply to the role of teaching and learning:

"Teaching shouldn't be about filling the bucket, instead it should aim to ignite the fire".

In this regard, the technology has to be seamless. You must consider which digital technology will be effective in 'lighting the fire'. The scenarios in this chapter will explore the features in different technologies and demonstrate examples of how to use them in the art of teaching.

Scenario 1

Ms Zeta, the language teacher has 68 learners studying a foreign language. She was going to be away from her students during their revision period and she wanted to ensure they'd excel in their forthcoming examinations. The only digital technologies that were available to all 68 learners were mobile/cell phones and WhatsApp because it was freely available to all her learners.

This is what Ms Zeta did as the teacher. She:

- set up a WhatsApp group for her 68 learners
- wrote a chat message explaining that she'd be away and how she would still guide them through the revision period
- uploaded a worksheet covering the main points they needed to revise
- recorded a voice message to explain two of the most difficult concepts they could expect in their examinations
- sent a link to practice exam papers from the internet

This is what her learners did using their WhatsApp messages from their teacher, Ms Zeta. They:

- joined the WhatsApp group
- read the messages
- listened to the voice recordings
- sent questions about sections of the work, they were still unsure about

Discussion

This scenario is very similar to what could have been happening in a traditional classroom setting with the teacher being present and together with all the learners. However, by using the WhatsApp group messaging system, all 68 learners and the teacher did not need to be together in the same location. They also did not need to be doing the teaching and learning all at the same time. This teaching and learning experience was remote and asynchronous in nature. The main focus, however, is however, how effective is this mode of teaching and learning.

Learners benefit from other learners comments, questions and answers besides those of the teacher. They can work when they are ready to learn wherever they are and at a time that suits them best. Often learners may have other distractions being in a classroom all at the same times, such as, not feeling well at that time, being hungry or needing to go to the bathroom. The relationship dynamics between young people can at best be complicated and not ideal to focus on serious learning content at times. Asynchronous and remote learning is not without its drawbacks but could have more benefits for effective learning.

Task 1
Look at the file, 'A Brainstorm of the Things I do with my Digital Device' seen below. Consider the items on the list and add anything else that might have been missed out.

Brainstorm of things I do with my device

- Place the items in the list into the following 2 categories: Those that require you to be connected to the internet and those that do not need to be connected to the internet. The reason is that internet connectivity may be an issue at times, and it is important to consider using digital devices that can still function without being connected to the internet.
- Refer to these online resources mentioned below and create guide sheets on how to:
    - create audio clips using your mobile phone
    - link a YouTube video to an electronic worksheet
    - use WhatsApp 'voice-overs' and send it to a whole group of students
    - conduct a group discussion in WhatsApp

Resources

https://www.wikihow.com/Category:Computers-and-Electronics

https://faq.whatsapp.com/

Discussion

The act of creating a guide sheet for others to follow demands that the creator of the guidebook should know how to do those tasks very well before they'd be able to instruct others about how to do them. This is a good way to 'light the fire' of your learners because it is an active form of learning. They are engaging with the learning material rather than the teacher just informing them of how to do it in a passive learning style. The teacher is now facilitating the learning process whilst the learner is actively making sense of the learning material.

The teacher is also suggesting possible links to sources of material online (OER) so that learners are not distracted by having to search through too many similar resources or sources of information that might not be suitable for their levels. The teacher would not necessarily be the 'fount of all knowledge', especially as technological advances come out at a fast and furious rate.

The output expected from learners could be as audio recordings, presentations, short video clips, paper printouts, information-rich sketches, screen recordings, etc.

Scenario 2

Mr Zunga had a class of students out on a field trip and they had electronic worksheets to complete during the field trip. All the students had tablets that were connected to the internet. Here are some of the tasks that students had to complete:

- collecting opinions from the public about recycling and waste management
- taking images of litter, waste, recycling, etc.
- working in groups to collect data and compile charts and graphs about the severity of a problem related to recycling, waste management or litter
- organise a community event that called for volunteers to help 'clean up the environment'.

Task 2

- Describe how students would go about completing their electronic worksheets. Mention the features on their tablets (cameras, voice recorders, etc.) that students are expected to use and the use of any applications programs (Word, PowerPoint, Spreadsheets, etc. or web-based applications.
- Create an assessment rubric for each of the tasks set by Mr Zungu using the Survey Monkey quiz tool.
- Suggest a method for Mr Zungu to share all the students tasks with each other.
- Describe the digital technologies you would use for a peer-assessment exercise for all the learners tasks.

Discussion

Learners could have use a variety of ways to collect information from the public using their digital devices, such as, preparing an online survey and sending the link to members of the public or they could approach members of the public and record their answers on the voice recorders or they could record short video clips where they receive the views of the public based on the questions they ask about waste management and recycling. Learners could also use their social media sites to elicit some views from the public on their study topic. The trend nowadays is to ensure that learners assure contributors that they would respect their data and privacy once they received their contributions.

Mr Zungu could use an online tool, called, TRELLO where learners could place their work and everyone in the group can see their efforts. This is a good tool for collaborative working styles. It also helps to improve the standard as students compete to make their successive efforts better than their peer's efforts.

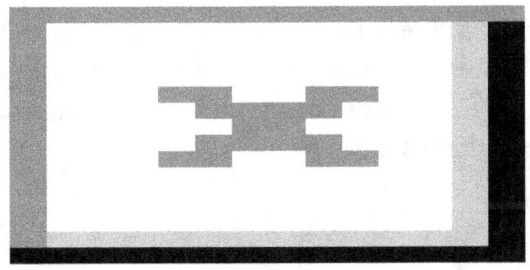

TRELLO – An example of a collaborative working e-space

Peer-assessments can be carried out by using the rubrics that can be created in a word processor and simply shared via email, WhatsApp, or the school's VLE (Blackboard, TRELLO, etc.). The feedback that peers give to each other will be argued and debated with each other on their own levels and may be better received from a peer than a teacher. A teacher may have to settle disputes arising from a misunderstanding of either the learning material or the assessment rubric itself. This is how true learning occurs.

Most learners are quite good at taking images but here is another teaching and learning opportunity for about the appropriateness and the platform to present the images so they have the desired effect. Would social media be a suitable digital technology to organise a community event to address some of the issues of waste management and recycling? Preparing such an event would require a lot of collaboration between many different individuals and learners should be guided very carefully through such efforts.

Scenario 3

Mr Jambo ran an after-school nature club. He asked each learner to find a small animal, insect, amphibian, aquatic or plant and flower for their project. Each student had to use their camera to record as much about the topic as they could. He suggested they use time-lapse photography and QR codes so that he could link each learner's project to the club's website. How would this come together?

Each student would/could:

- select their topic, e.g. a tadpole that would turn into a frog or a caterpillar that would turn into a butterfly

- take photographs of their topic as time goes perhaps, once a day, every two days, or every week, etc. depending on the topic
- look up more information about the life-cycle of their topic
- prepare a file storing all the information and images as they find out more about it
- arrange the information onto a document or presentation
- prepare the QR code for the document or presentation
- send the QR code to the teacher

The teacher would/could:

- ensure that learners knew how to use their cameras
- ensure that learners have chosen suitable topics according to their ability levels
- ensure that learners knew how to create and use QR codes
- checked on the weekly progress of each learner
- collated all the QR codes from all the learners
- prepared the nature club's website
- uploaded each learner's project onto the website
- made the website link available to all the learners for their acceptance before publishing it for a broader audience.

Discussion

This exciting type of teaching and learning experience could be used for many different subjects. The time-lapse photography could be used in Science, e.g. watching how a fruit decays over time or how the weather changes over a period of time.

The use of QR codes is exciting for students because generally, they love interacting with digital technologies and learning new things at the same time. My favourite use of QR codes is creating mathematical problems on one QR code and the solutions on another QR code. Print out the QR codes and jumble them up. Hand out one QR code to each learner and ask them to match the problem with the correct answer.

Task 3
Evaluate the usefulness of the device's features for teaching and learning for each of the lessons above. List their advantages and their disadvantages.

Scenario 4

Learners mostly love songs and music and what a good way to engage learners in a fun way to learn about almost anything. Mrs Hoohaa asked learners to create a song about the capital cities of a list of countries. Each learner or group had record their songs for just two capital cities of two countries from the list. The song had to include some interesting facts about each city, such as, France has the Eiffel Tower in Paris, etc.

Students could/would:

- find out facts about their two countries and their capital cities
- create the lyrics for their song
- create a suitable tune for their song
- record their song
- send the sound file of their song to their teacher

The teacher should/could:

- suggest links to suitable resources
- give learners an exemplar song
- suggest a suitable file type to store the song
- upload learners songs onto the audio editing platform
- edit and stitch their songs together if suitable
- playback the class effort for all to listen and comment on
- suggest students use their combined song to learn about the capital cities of all the countries on the list.

Discussion

Recording audio content on a mobile phone or tablet is commonplace today. Most young learners already know how to record audio content. Perhaps they should be reminded how to store their audio clips using sensible file names and in appropriate folders on their devices. In these instances the technology itself has to be a seamless aspect of the learning process. Teachers and learners should not allow the use and operation of the technology to distract them from the main learning content at hand.

Examples of good uses for audio recordings are perhaps where learners record their language orals or music learners send their musical renditions for auditions. Video recordings can be done using the device's camera. Video editing can be done using a different piece of software another time, if necessary.

Using online presentation applications, such as, PREZI, doesn't include using a device's audio or video recording features and as such, serves as an alternative types of task for learners to demonstrate their understanding and grasp of learning content.

Practical Task for Teachers

Catering for learners of different abilities is generally termed, differentiation. This can take several formats as explained in this video: https://www.youtube.com/watch?v=h7-D3gi2lL8

One example that springs to my mind, is when I gave my students the topic of 'Health and Safety in the Work Place'. Many learners began to sigh and though it was a boring topic until I mentioned that they'd need to use their mobile phones to find evidence of breaches to this policy around the school. Needless to say, the school management were not very happy with the outcome, but it resulted in many unexpected outcomes. The school became a safer place to work at, learners knew exactly what the policy contained and they learned how to use the functionality within their devices to engage with the learning content and topic. Learners also presented their findings to their peers, thereby honing their oracy skills.

- Consider how students who are:
    - less confident with the subject matter can be supported by using a digital technology
    - more confident with the subject matter be challenged by using a digital technology
- Describe what strategy you would use next time to enhance the teaching and learning experience using other features/functionalities within digital technologies even more.

Using APPs in Teaching and Learning

3

In this chapter, you will:

- explore a range of APPs relevant to your subject/discipline
- integrate a range of APPs in a teaching and learning episode
- review its effectiveness as a tool that could enhance the teaching and learning experience
- recommend and justify using specific APPs for your subject/discipline

Introduction

APPs are short programs that have to be downloaded onto your device, mostly via your device's APP Store. Some APPs are free, but not all of them are free, so be careful when signing up for APPs.

APPs are amazingly hypnotic tools for teaching and learning specific sections of a broader course. The content in APPs do not necessarily include all the content required by most countries state-approved syllabi or curricula and in this regard, APPs should be used as an aid to teaching and learning.

It can be tempting to use too much of an APP in the teaching and learning of a specific concept or skill, that you have to be absolutely sure that the original learning outcomes can be met by its use.

If you use an APP to learn a new language (for example), only the learner might have access to their own progress and a teacher will not know how much progress each student will have made during a specified learning period. However, as the teacher, you would have to think of ways to integrate the use of the APP within a section of your broader scheme of work for learning.

I'll refer to the language learning APP, 'Duolingo'. The style it uses is to allow you to memorise increasingly more new vocabulary and sentences as you progress through the 'lessons'. This style of teaching may not be how you teach the language. You may be teaching a section on 'nouns' or 'verbs' as separate lessons. The Duolingo APP wouldn't be of much help for such lessons. However, you could use the APP to reinforce what you've taught in your lesson on 'nouns' or 'verbs' and to allow your students to practise what they've learned in your lesson.

Scenario 1

Mr Thomas uses the APP, "Maths Champions" at the start of each Maths lesson to get students ready for each lesson. He could add each learner's name and then either sets them a challenge or a training session. He doesn't feel that he needs to add their names as this part of the lesson doesn't require tracking, it's just a warm up session before the main lesson topic.

Discussion

Although this APP is quite useful and requires very little preparation from the teacher's side, you should think about other more appropriate APPs because it's not always best practice to use the same type of starter and it may be more appropriate to use a topic that is related to the one you are teaching in that lesson.

Task 1

Refer to the video above which can be found on here: https://www.youtube.com/watch?v=vTDh-h7Xti4&feature=emb_logo . Make short notes about how you may be able to use any of the 10 APPs mentioned in the video. Perhaps, create accounts in each of these APPs and experiment with different classes to see which works best for your teaching cohort.

Scenario 2

Ms Xioux teaches in an online school and has created an account in Google Classroom and has created a board for each of her four classes. She has created Topics for each of the sections in the curriculum or syllabus. She uses Google MEET to connect via video link to her learners and the learning materials are uploaded into Google Drive. She uses the chat section to keep the explanations and discussions going throughout her lesson and opens a whiteboard as if she is writing on a writing board in her traditional classroom. The lesson is recorded for learners to go over any aspect they may not have understood at the time it was being done.

Discussion

This mode for distance teaching and learning has many advantages and might take a bit of practice before everyone is comfortable with it. The recording can help learners to go over sections of the work they struggle with at their own pace. It's also easier to forget what had been said in a traditional lesson, whereas a recorded lesson can be replayed over and over again when necessary.

Learners do not have to be present during the teaching session and can access the lesson afterwards but will obviously not get the same benefit as those learners who do attend the online lesson in real time. The big idea behind meeting online face-to-face is that learners can interact with their teacher and other learners as they would if they were in the same classroom at the same time. They could ask questions about a section they needed clarification on at the time when the query arises just as they could ask their teacher in a traditional classroom setting.

Task 2

Create an account to use 'Padlet'. Pose a question that your students could answer at the end of one of your lessons. Check their responses for next steps.

Create an account in 'Socrative' and find a quiz that is relevant to a topic you are teaching or create your own quiz. Use the 'Socrative' quiz on your students and analyse a breakdown of their results.

Discussion

A reasonable question to ask now is: "Can I use and APP within another APP?" If we call, Google Classroom and APP, then I suppose the answer is YES because APPs, such as, Padlet and Socrative, can still be used while you communicate with your learners through Google Meet. Remember to aim towards having the digital technologies that you use as seamless as possible. The main objectives should be taught using the most effective and least disruptive or distractive tools available to you and your learners. If it's more effective to use a hardcopy textbook, then use that rather than waste time and effort searching for the same information on the World Wide Web. Its wise to adopt a common sense approach.

Scenario 3

Mr Chikasa runs his own YouTube channel (https://www.youtube.com/c/EvansChikasa/featured) showing videos that help learners to answer past exam papers for the subject he teaches, ICT. He answers each question by demonstrating the practical ICT skills needed to answer a typical exam paper. He also shows learners where and how marks are assigned for each part of a question.

Task 3

Create an account with www.youtube.com in order to create your own YouTube channel for demonstrating to your learners how to solve a particular type of problem or to do something relevant to your subject.

Create a 'Test Video' and perhaps experiment with doing a 'Live streaming event' as well. Make it available to the public instead of private and see what response you get, if any at all. You may wish to direct your learners to your 'Test Video' and ask for their feedback so that you can judge whether it would be a resource you could/would like to create or use.

Create a 2-page leaflet to guide fellow colleagues on the benefits and pitfalls of using their own YouTube channel as a teacher/instructor/trainer.

Discussion

YouTube allows anyone to run their own YouTube channel. You can create videos and even do live streaming to share your teaching resources with anyone, including your students and colleagues.

Learners simply subscribe to his YouTube channel and watch his videos. This facility could be done by any teacher. However, why would you want to re-invent the wheel in this regard. If someone else has already created a revision resource like this, then perhaps refer your learners to their link. However, there may not be anything suitable for your learners and you could possibly create your own YouTube channel with the added bonus that if it gets more than a certain number of views, you could possibly earn some extra money along the way. Learners can benefit from subscribing to channels, such as, Mr Chikhasa's because he goes through ways to answer exam paper questions thoroughly and it can be replayed over and over again. Learners will know more precisely which sections of their work they will need to revise more thoroughly or to seek more assistance from their teachers.

Scenario 4

Miss Barrow uses Plickers for each of her five Biology classes assessments. She likes to assess her learners bit by bit throughout her lesson, so she teaches a small section or a concept and then sets a question or a set of questions in Plickers. Learners use the Plickers cards that she has prepared for them to scan onto her phone or iPad APP and that gives her immediate feedback as to how well her learners have grasped the section or concept that she just taught. Take a look here: https://www.plickers.com/missbarrow/The-Carbon-Cycle-3588

Discussion

Assessment can take many different formats but the Plickers APP is one that injects a lot of fun into the process of assessment. You could use an existing Plicker in your lesson or you could edit an existing one by adding some of your own questions and deleting those that are not relevant to your lesson. Alternatively, you may wish to create your own Plicker from scratch.

This is what the teacher does. They:

- create their own questions or select from the existing bank of questions
- project and display the questions on the board
- hand out the answer options cards to learners
- track the results after learners have scanned their answer options on their iPhone or iPad Plickers APP

This is what learners do. They:

- read the question displayed on the board
- make their choice for the answer, either A, B, C or D

- scan their card, holding it the right way up to scan their chosen letter option on the teachers iPhone or iPad

Task 4

Sign up for a 'Plickers' account and set an assessment or short test for your students. Conduct the assessment or short test using 'Plickers' and then analyse the results of your students.

Practical Task for Teachers

Create a lesson plan for a relevant section of your teaching programme. Try to use as many APPs as you possibly can.

Consider the effectiveness of each APP and gauge the levels of engagement of your learners during your use each APP.

Using Video effectively in Teaching and Learning

4

In this chapter, we will:

- explore how to use video more effectively in a variety of ways to enhance the teaching and learning experience
- prepare and deliver parts of lessons using each of the different ways to use video in teaching and learning
- evaluate the effectiveness of each strategy
- reflect on the most appropriate use of video for specific pedagogies

Introduction

The effective use of video in the teaching and learning experience is one of the most under-utilised teaching strategies, despite video technology being around for quite a long while now. The most common use of videos in lessons is where a video is simply switched on and left on for students to watch it from the start to the finish and then they have to answer some questions on a worksheet. Whilst this method of teaching (you might wish to question how passive such 'teaching' might be considered to be) is the most common, it isn't suitable for all age groups of learners and for all ability groups of learners. This method should be largely avoided because learners soon lose interest, especially if the video is longer then their concentration spans.

You have many options to approach the use of videos as part of your teaching and learning plans. However, video editing can become a very useful tool if you wish to use very specific parts of an existing video footage. On the other hand, you could simply stop and start a video so that it plays only the sections you wish to use for your lesson. You could also create your own video clip using the movie-making capability within most digital devices.

If you wish to raise the levels of engagement for your learners, you could always ask them to create short video clips based on the learning materials. Learners enjoy working in small groups and they enjoy watching each other's videos and making suggestions on how to improve the output of it in a fun way.

Scenario 1

Mr Vidal is teaching a languages class in creative writing and uses a short video clip with a simple story line. However, he keeps the sound off during the watching of the video and then asks his learners to write down what they believe happened in this story just from watching it without the sound. Once the learners have shared their versions of the story in the video, Mr Vidal played the video again but this time, with the sound turned up so that his learners could compare their versions with the version in the video. Mr Vidal asked learners to consider why they may have imagined differently from the events in the video.

Discussion

This is a good example of active learning because learners are not passively watching everything and have to imagine throughout what could be happening. They have to think of several possibilities. Whether they are correct or incorrect is not the point. The point of such an exercise is to give learners the opportunity to think through many possibilities and imagine creatively. Depending on the capability of the learners, you might wish to ask them a few questions to lead them towards a particular idea. There is also opportunities for discussions about differences of opinions and much more.

Task 1

Using Mr Vidal's lesson information, write up 4 to 5 possible:

- lesson objectives (aims for your lesson)
- learning outcomes (things your learners must produce by the end of this lesson)

Scenario 2

Mrs Malthi is also teaching a languages class and she uses the same video that Mr Vidal used in the scenario mentioned above. However, she asks learners to turn their chairs so that they face towards the back of the classroom and they cannot see the video but only hear the sound track. She asks her learners to draw a storyboard to depict what they believe is happening in this video. She provides a worksheet with 10 empty boxes for them to draw in and to write a few lines about each drawing beneath it. Learners swap their storyboards afterwards and discuss their different thinking about what they believe could be in the video. Mrs Malthi then shows the learners the whole video and leads a discussion about any differences in some learners drawings.

Discussion

The use of video in this way has to be very carefully planned so that the lesson objectives and expected learning outcomes are achieved by the end of the lesson. It may not be obvious initially, but the use of video in this way can apply to most subjects as can be seen in the following examples:

- use of case studies (Humanities subjects, such as Geography, History, etc.)
- processes e.g. explain how blood flows through the heart and lungs (Science)
- solving mathematical problems, steps to follow
- how to do some practical tasks, e.g. seeing how to carry out an Art technique or designing something (Design or Food Technology, Art)

Task 2

Complete the video storyline – watch a video (with or without the sound) up to a critical point in the storyline of the video and stop. Ask learners to complete the storyline with their own ending to the story. Ask learners to share and discuss their different storyline endings.

Scenario 3

A Computer Science teacher wants to use a video to teach his learners about a bubble sort algorithm. Here are two different videos found on YouTube.

https://www.youtube.com/watch?v=lyZQPjUT5B4 and
https://www.youtube.com/watch?v=yIQuKSwPlro

Discussion

He then asked his learners to use the same concept from the video to complete the table below by tracing the operation of bubble sort on the following list of numbers: 4, 7, 2, 5, 6.

This teacher is faced with selecting the best way to teach learners how to do a bubble sort algorithm and to ensure that learners will be able to understand the concept being demonstrated in the teaching materials and then be able to complete the table correctly as proof that his learners have achieved the learning outcome for this lesson objective.

Task 3

Provide commentary by using these questions to ponder the effectiveness of each video:

- would the dancing in the first video distract the learners from grasping the whole idea?
- would the first video take too long for the whole concept to be spotted by the learners?
- would learners be able to make the link in the underlying idea/concept from dancing to sorting numbers in a table?
- would the second video suit my learners better than the first video because it is also using numbers?
- should the teacher perhaps consider using aspects of both videos to help the learners make the link between the dance and the numbers?

Questions to ponder about how to use the videos:

- might it be more engaging or effective to ask learners to create their own dance for a bubble sort as proof that they understand how to do a bubble sort algorithm?
- might learners be asked to think of a different scenario to apply a bubble sort algorithm?
- might you ask learners to create their own short video clips showing how to do a bubble sort algorithm?

Scenario 4

Mrs Butters, the foreign languages teacher, is using a video that speaks in a foreign language. She provides her learners with the video already placed in video-editing software. She asks her learners to provide sub-titles in English for the video by adding text boxes over each of the video panes.

Discussion

This way of engaging with the learning material could also be achieved in other ways and would only work if all learners had access to video editing software on their digital devices. Learners would still have to identify each of the foreign words and phrases and then translate them before adding them as sub-titles on each video frame. The idea is that learners have an opportunity to translate from one language into another in a real-life scenario. They would learn much more than simply new vocabulary and phrases of a different language. There is also the possibility of using electronic translators.

Task 4

- List a few different ways that you could use sub-titles in video clips in a teaching and learning experience.
- Write down the lesson objectives for each of the different ways you have listed in question (a) above.
- Write down the learning outcomes for each of the different ways you have listed in question (a) above.
- Design a possible assessment rubric for each of the different ways you have listed in question (a) above.

Scenario 5

Ms Ramsulah has asked her learners to create videos where they act as weather presenters for a local TV station. Learners should use their mobile phone video recorders or use pictures taken with their mobile phones and then they should add voice-overs or text-overs.

Discussion

In this teaching and learning experience, the teacher would have to:

- prepare clear assessment criteria so that all learners know how their videos would be assessed before they begin
- teach the relevant weather-related learning material before learners begin their task
- ensure learners have all the resources for creating the video, such as, video editing software
- ensure that learners prepare a script to follow during their presentations
- ensure learners work in small workable teams to achieve a reasonable quality of video presentations

Learners will have to:

- research by watching how professional weather presenters act on their local TV stations
- learn about the weather phenomena they intend to cover in their weather presentations
- prepare a script to act out during the video recording of the weather presentation
- check that the presentation meets all the assessment criteria before finalising it

Practical Tasks for Teachers

Task A

Select a video either from an online source or any other appropriate video that you have available for a teaching and learning event.

Compile a few different lesson plans using as many of the different ways you can use video in teaching and learning.

Task B

Create an account in the APP, 'EdPuzzle', and insert 3 appropriate questions at key points in a video you would be using in your lesson.

Task C

Create a short questionnaire that evaluates the benefits and drawbacks of your use of video in your teaching and learning for any ONE of the lessons you planned in Task A.

Task D

Recommend useful videos that other teachers in your discipline/subject specialism could use.

Using Online Platforms in Teaching and Learning

5

In this chapter, you will:

- explore the possibilities of using online teaching and learning platforms, including virtual learning environments (VLE)
- select an appropriate online teaching and learning platform or VLE technology and use it to create a teaching and learning event
- review the usefulness of your chosen online teaching and learning platform or VLE tool as a means to conduct a teaching and learning session
- analyse feedback from your students and make recommendations for further improvements to the teaching and learning experience

Introduction

Online teaching and learning platforms including VLEs have become a fairly common tool being used in face-to-face communication when participants are located remotely. Online teaching and learning platforms and VLE technologies include online tools that combine the use of voice and video technologies which means that participants can talk and see each other as if they are in the next room, despite possibly being halfway across the world. VLEs also include storage for documents, videos, tools for collaborative tasks and some also have built-in assessment tools. Examples of VLEs include Google Classroom, MOODLE, BlackBoard, FutureLearn, Schoology, Edmodo, ManageBac, ClassIn, etc.

Some online tools have a wide appeal for teaching and learning yet cannot be clearly categorised because each offers very different functionalities and applications for teaching and learning. Examples include, Wolfram Alpha, TRELLO, Survey Monkey, Khan Academy, QR codes, electronic translators, MOOCs, online tutorials, such as, w3Schools, the gameficiation of learning, etc.

The implications for teaching and learning are enormous. However, there are arguments for and against its use, especially as the Digital Divide continues to widen the inequality gap regarding access to a relevant education for all. Despite the many free platforms available, alternative free platforms offer much of the essential functionalities found in the more comprehensive and substantial VLEs. Platforms, such as, MS Teams, ZOOM, SKYPE, WhatsApp can also be fairly accessible for those with a minimum of internet access and much less data available to them. These alternative free online platforms do offer a great deal of flexibility and a teacher could adapt particular digital technologies to achieve similar learning outcomes for their learners.

In the first two scenarios for this chapter, you will explore using a VLE in on different ways. And the remaining scenarios will be dedicated to alternative online platforms that can be used in teaching and learning but may also be applicable in other disciplines, e.g. Survey Monkey is a quiz making tool and is often used in businesses seeking feedback from customers, marketing purposes or making quizzes for fun.

Scenario 1

Ms Maitland is working remotely and teaching a group of learners in a faraway country. They all sign in to their VLE accounts at the specific time for their lesson. Learners find all the learning materials for the lesson in the storage area and open them in readiness for the lesson. They also upload any previous homework and assignments that were due before this lesson. When everyone is connected to the VLE portal, Ms Maitland starts the lesson by explaining which resources learners should refer to and she speaks to them using her webcam and microphone. Learners do the same and can talk to her in the same way. Although they may not all be together in the same location each learner can also talk to each other in the same way that a WhatsApp group can communicate with each other by replying to any individual in that group using the chat facility of the VLE. At times Ms Maitland shares her screen and underlines text or writes on her writing board and all learners can see what she has done.

Scenario 2

Mr Isaacs has been using a VLE in his everyday teaching at the Belhar School, where learners and teachers are physically present in the same location. He uploads assignments, homework, tests, video links, etc. prior to the lesson and reminds his learners to upload their homework and completed assignments onto the VLE. The school has a policy to encourage using a paperless teaching and learning environment. There are no printers at this school, however, they still use textbooks for most subjects. This way of mixing traditional teaching styles with computer-aided teaching and learning is referred to as the blended approach.

Discussion

Scenario 1 represents a typical remote online lesson using a typical VLE. However, there can be some disadvantages faced either by the teacher or the learners. Wouldn't it be better if learners had two screens or two devices to work from, one for the discussion with the teacher or other learners and another screen or device to look at resources that have been sent by the teacher. (Similar to looking at a teacher or other learners and also at the worksheet or textbook during a traditional face-to-face lesson).

A distinct advantage is that there is less opportunity for learners to be distracted by other learners with behavioural issues, so there is more time for a more focused learning experience. Teachers can progress more speedily through the learning content in such cases.

Some VLEs record and store each lesson and learners who were absent at the time of online lesson can view the lesson asynchronously and at their own pace. The advantage is that all learners can access their lessons at any time in the future and replay any section of the lesson they may wish to use revision purposes. In cases where the actual lessons were not recorded, learners would still have access to the learning materials used during those lessons.

Scenario 2 focuses on a blended approach which can be flexible and occur in many different ways. It offers teachers time to adopt as much or as little as they are comfortable or confident with when they choose to use digital technologies in their teaching practice.

A benefit for teachers is that once learning material is uploaded and stored, it is available for as long as the teacher gives their learners access to it. If either the teacher or the learner is absent from school, the material is still accessible online and there would be less disruption to the flow of working provided learners are guided to what they are required to do next.

Additional ways to use VLE to create exciting teaching and learning experiences includes:

- working in a subject-specific online learning platform, such as, Khan Academy. Here you can create a Class and track your learners progress as they progress through the learning material. You could use this online platform to differentiate the learning opportunities for a class with a large number of students. By identifying students who may have slowed down their learning (seen from their progress in Khan Academy's tracking system), you may intervene and guide these learners by using a different method to convey these more difficult concepts to them. Typically, you will be differentiating the teaching in this way.
- working remotely and working through a textbook together with your learners. This way may/may not be the most effective way to conduct a teaching or learning experience. On the other hand, some textbooks may be excellent and could possibly lend itself to some creative ways to facilitate the learning process.

- working remotely and using online videos where learners can either watch from your VLE platform or use their own digital devices to click on video links you would have uploaded prior to the lesson or sent via the chat window.
- creating quizzes and online assessments could similarly be sent or uploaded in the same way, such as, a short test could be created on Survey Monkey and the link sent to learners. You could then view the results as soon as all the learners have completed it but in another window not through the VLE platform.
- combining the use of APPs by sending web links via the chat section of the VLE platform and then tracking or monitoring learners' progress in another window other than that of the VLE platform.

Practical Tasks for Teachers

Task A

- Research the acronym, 'VLE', and list all the examples that you are familiar with so far.
- Make short notes on the things you need to conduct an online or VLE teaching experience for both the teacher/instructor/trainer and the students.

Task B

- Find a relevant video resource online for information on how to set up your lesson using ZOOM.

- Using the topic, 'The future of environmentally-friendly teaching and learning experience', create a lesson plan that reflects this topic whilst being conducted through ZOOM.

Task C

- Share your lesson with up to 5 other teachers/instructors/trainers and ask for their feedback on the effectiveness of using ZOOM as a teaching platform. Write replies to each of them and save these and their feedback in your portfolio for future reflection and improvement purposes.
- Improve your lesson plan by incorporating the feedback from your colleagues and save in your portfolio.

Task D

- Conduct your lesson you created earlier with your students.

- Create an evaluation that elicits feedback from your students to find out how well they have learnt using ZOOM.
- Tweak your lesson plan once more according to the feedback received from your students and save it in your portfolio as well.

Scenario 3

Mrs Jordan teaches Computer Studies and uses many different online platforms, portals and tools. In her lessons she is teaching the topic animation and she has to teach her learners how to code and how to use animation software to create animations. Here is an excerpt from her lesson plan for this topic:

Starter: What is animation?

Mrs Jordan uses the video from YouTube to teach her learners about what animation is: https://www.youtube.com/watch?v=dGGU4rAkShE

She ask students to write down in their own words what they think animation is. A whole class discussion ensues.

Main Teaching Section: Using software to create animation

Shes uses the software available to learners from the school, Adobe Animate. She demonstrates by explaining what some of the key features of the software can do and gives learners time to practice using the software's features. She sets a task for them to create different types of animations and they share their efforts by giving feedback in a peer-assessment exercise at some point in the lesson.

Discussion

However, there are several free versions of software for teaching and learning how to create animations, such as, Animaker, Blender, K-3D, Open Toonz, Powtoon, Pencil2D Animation, Skykz, Plastic Animation Paper, etc.

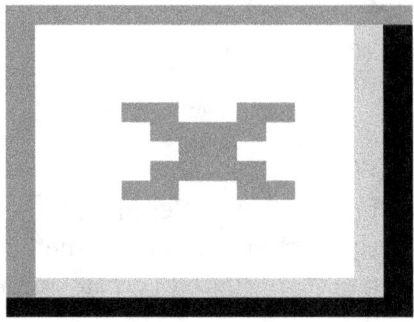

Examples of animation software

taken from a Google search on 25/10/2020

Main Teaching Section: How to code animations

Mrs Jordan also refers her learners to Khan Academy so that learners can learn how to code animations on their own. She uses what is called the 'Flipped Classroom' method here because she will only intervene in the learning process for those learners who are struggling with particular sections of the lessons provided on Khan Academy's video and tutorial-based learning events.

Mrs Jordan is able to track each learners progress by signing up for her own account and inviting her learners to participate in the tasks she sets them. Here is the link to the section on coding animations that she uses for her learners:

Once the learners have attempted some of the tasks and challenges set in the programme Mrs Jordan can track their progress and know when to step in to assist any learner on a more one-to-one basis by using the tracking facility available on Khan Academy.

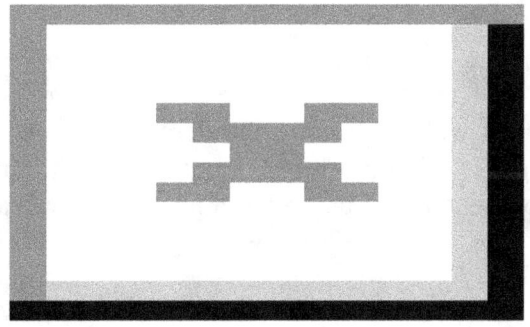

Example of how to track learners' progress

Plenary: End-of-topic Assessment

Usually, an end-of-topic assessment takes the form of a short test of some sort. The question with the way that this section or topic has been taught and learned is, 'would learners really benefit from further testing?' Assessment has been built into each sub-section of this topic with challenges being set after each demonstration section and if learners couldn't complete the tasks and challenges, they wouldn't have been able to complete this topic. Besides, since the teacher had been tracking each learners progress through this learning episode, she would have had to intervene and assist any learners who were less confident than others.

However, should a typical test be required (in case the school's assessment policy demands a standardised type of test for all end-of-topic sections), there are other interesting online assessment possibilities, which will be explored in the next scenario.

Task 3

- Create a 3-part lesson plan using different digital technologies for your own subject.
- Comment of how you believe your lesson objectives will be met by using the chosen digital technologies.

- Predict any drawbacks that may arise during the delivery of your lesson using the chosen digital technologies.

Scenario 4

Ms Dennis wants to test how well her learners understand the concept of measuring volume. She uses an online learning platform, StudyLadder that has a built-in test after she used some of the 'printable resources' as revision notes for her learners.

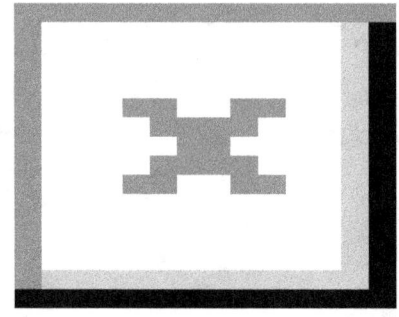

Online assessments freely available on www.studyladder.com

Mr Beaton teaches the same topic to his learners but he uses the BBC Bitesize resources found here: https://www.bbc.co.uk/bitesize/topics/zrwkgwx/articles/z3jrxfr

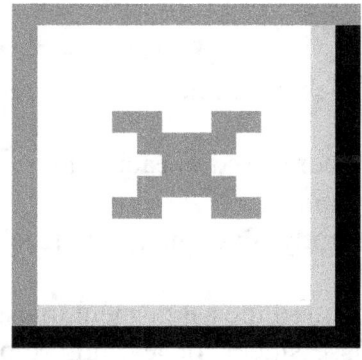

Online tests or assessments on www.bbc.co.uk

Discussion

Both teachers do not need to waste time to create new tests when there are already so many excellent tests and assessments that they could use and that are freely available online.

These tests get marked automatically and the results can be viewed by the respective teachers. However, the result alone doesn't help either the teacher or the learner because the learner wouldn't know precisely what section he didn't understand properly and the teacher would not know how well each learner understood the concept. These types of tests are not always diagnostic enough so that the learner benefits from doing the assessment in some way. These summative types of assessment tools are more beneficial for administrators and parents. When assessment questions inform administrators or parents only and have little benefit in extending the learning for learners, then it is referred to as 'assessment-of-learning' (AoL). Examples are tests results that give no feedback to learners and often compares learners results against each others. They usually summative, such as, end-of-unit tests or examinations. On the other hand, 'assessment-for-learning' (AfL) benefits learners by its useful feedback, comments and suggestions on how to improve. Examples are the marked work whereby teachers make comments, suggestions on how to solve problems and generally acknowledge a learner's good efforts.

Task 4

- Use Survey Monkey to create multiple-choice-questions to assess your learners based on a recent section of the curriculum that you've taught.
- Create a graph to show you:

    - an analysis of how well each question was answered by your group of learners
    - each learner's total result compared to each other
- Based on analysis in your graphs/charts above:
    - what would you do differently if you had to re-teach that section of the curriculum?
    - how could you have structured the assessment questions so that the assessment was of the 'ássessment-for-learning' type in favour of it being of the ássessment-of-learning' type?

Scenario 5

Mr Mathews is a Maths teacher who uses a relatively unorthodox method in his teaching. He is teaching the topic on how to solve linear equations and he goes through the steps in class by talking through the steps with the whole class. However, he knows that not all his learners have followed or paid attention enough to master this section of Maths. In order to overcome this problem, he gives is learners several linear equation problems to solve. However, he asks his students to use Wolfram Alpha to solve all their problems and informs them that he is not interested in their final answers but that they need to learn the steps for solving the linear equation problems that he gave them. By repeating the steps, he knows that will consolidate the teaching he tried to do verbally during his lesson that day.

Discussion

Wolfram Alpha is a computational engine that can solve mathematical, and science computational problems. Here is an example of the window for solving the linear equations:

https://www.wolframalpha.com/input/?i=solution+to+2x+-+3+%3D+1&lk=3

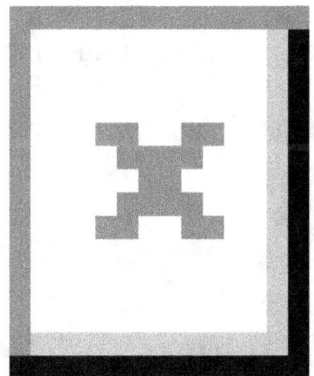

Computational Engine Wolfram Alpha

Wolfram Alpha can be used for many different subjects as seen in the image below:

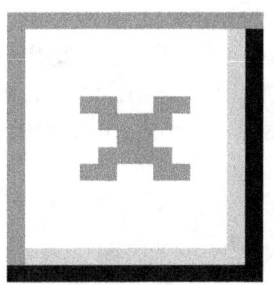

Choice of subjects relevant for using Wolfram Alpha

Task 5

- Create a quiz whereby learners should use Wolfram Alpha to find the answers within a specified timeframe for your subject. Example, in History, you could ask them to find out 3 facts about the 44th President of the United States of America within 2 or 3 minutes or in Chemistry to balance chemical equations within 2 or 3 minutes.
- Comment on how effective this method of teaching and learning might/might not be.
- Explain how could Wolfram Alpha be incorporated into your teaching style?

Using Applications Software in Teaching and Learning

6

In this chapter, you will:

- explore some of the functionality within applications software
- apply these functionalities to teaching and learning contexts

Introduction

Applications software includes our everyday software that we use, such as, word processors, spreadsheets, presentations, drawings, video editing, etc. It is safe to assume that the majority of us use the minimum functionality with each of these software packages. In this chapter, I will explore a few uncommon functionalities that could be useful in the teaching and learning context. For the purposes of this book, I'll be using the MS Office suite software in the examples explained.

Scenario 1

Mrs James often uses MS Word to create text for stories that she wants to use in her lessons because she feels that by writing her stories it doesn't come out the way she wants when compared to her telling the story. She uses the 'voice-to-text' (Dictate) feature of MS Word. This was recorded in MS Word: *By pressing the dictate button on the top right hand corner of the Microsoft window menu bar you will notice the red button appears when you click on it that means you are recording everything until you click on the button again and it will then switch off.*

Discussion

Take note that your grammar may not be perfectly correct when you speak so you may have to make a few edits to your written version before sharing it with others. However, the benefit is how quickly you can fill a page by using the 'voice-to-text' tool.

Within a teaching and learning context, you could ask learners to tell their best story using this 'voice-to-text' tool and then ask them to swap with a peer to make some grammatical corrections depending on what your lesson is about.

Another way to use the 'voice-to-text' tool is to have an assessment/text as an e-document and ask learners to speak their answers using the 'Dictate' button. The only problem with this is that other learners may hear each others answers. However, you could ask learners to keep their headphones on.

Task 1

Create a worksheet where learners giving learners the choice of using their 'voice-to-text' tool to complete the worksheet before submitting it to you electronically. Bearing in mind that not all learners may wish to use the tool, see if you can detect which learners did use it and which ones didn't use it. There should be clues visible in the script. What are the reasons why there is a difference between their written text and their 'voice-to-text' scripts?

Scenario 2

A few of Ms Taylor's learners had used the grammar and spell checks before handing their essays in assessment. However, Ms Taylor still had to assess each learner's essay and he used 'Comments' to give feedback to his learners, making suggestions about better ways to re-phrase or more appropriate vocabulary to use. When the learners received their marked essays, they used 'Tracking' so that when they resubmitted their improved essays, Ms Taylor could see how their original essays had been amended. In the image below, You can see the comment inserted and the suggestions made for the learner by the teacher through the 'Comments' tool.

Using comments as part of an Assessment for Learning (AfL) exercise

In the image below, the text in red and underlined and the text with the scratch through line to delete it are all the changes made by the learner as a result of the suggestions made by the teacher in the image above, through the 'Tracking' tool.

Using Tracking to respond to earlier comments

Discussion

When writing in the language, English, you can use grammar and spell checks to ensure that your text is as grammatically correct and contains as few spelling errors as possible. A useful tool for improving grammar and spelling are software programs, such as, 'Grammarly', which is an add-in in the MS Office suite. Grammarly highlights and makes suggestions for your incorrect grammar in documents that you are writing. Similarly, Spellcheck, helps you to improve your spelling by underlining incorrectly spelt words and suggesting alternative corrections. However, they should be used with caution because homophones, could go undetected when you might have spelt a word correctly but in an incorrect context, e.g. in the sentence, "I am *board* of writing on the *bored*". Both italicised homophones are spelt correctly but in the wrong context. It should have been highlighted by both 'Grammarly' and the built-in 'Spell Check' but none of these errors were highlighted in the text above as errors. This means you have to have your work proofread as well.

When a teacher makes comments for learners to improve or amend their work (this is excellent as an AfL exercise) and learners respond by accepting their suggestions and using the 'Tracking' tool, you can see how the AfL works through as many revisions or iterations as may be necessary before a learner has demonstrated mastery of the learning material. This method also helps to personalise the learning through the 'conversation' that develops between the teacher and the learner.

Task 2

Create a task for learners to use the grammar and spellcheck tools. When you receive their task, use the comments tool to mark their work and ask them to correct their work based on your suggestions and request that they use the Track tool before the make any improvements to their work and resubmit their improvements to you thereafter.

Scenario 3

Mr Daniels prefers to use MS PowerPoint to display each part of his lesson on the whiteboard as he teaches. Below is an example of what each slide contains:

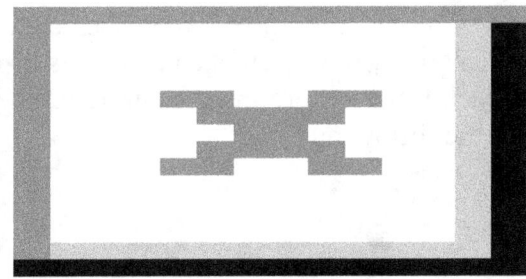

Using MS PowerPoint for pacing the delivery of your lesson

Discussion

The way that these presentation slides are used is a bit unconventional in that there may too much text on each slide but the big question for a teacher trying to teach using only the information on these slides, some worksheets, the short video and no textbooks is, "how effectively will my learners learn if I use my presentation slides in this lesson?" I would say that it is quite effective because learners listen to what the teacher is explaining for some of the time, they write some answers on their worksheets some of the time, they watch a short video and then draw a sketch and discuss with their peers some of the time.

The presentation lets the teacher control the pace of the lesson very effectively. By using as much text on each slide gives learners time to imbibe the concepts and time to look and listen simultaneously during the learning process.

Task 3

- Create a written lesson plan for the lesson portrayed in the presentation slides.
- Suggest ways to amend this presentation that might improve learner outcomes by the end of the lesson.
- If you were to use MS PowerPoint as a means to deliver your own subject's lessons, what would you do differently on your slides?

Scenario 4

Ms Hemmingway works in a large school and she has many different classes with a total of nearly 300 learners. It can become a nightmare when she has to write reports for all 300 learners because she has to remember how well each learner is or isn't doing in order to produce reports that reflect each learner's progress accurately. She uses a spreadsheet to make her job much easier in the following way:

MS Excel – conditional formatting to track learner progress

The green, yellow or red colours in the first few columns depicts how high or low each learner's scores are. The scores in blue is adjusted as a percentage of the whole cell, such that, if half of a cell is blue then the score is about 50%. In the cells with arrows pointing upwards, downwards or sideways depict good progress, a downward trend or a trend that represents the status quo. A formula is applied to the last column so that a relevant statement will appear based on the progress of each learner. The colours can easily be created by using the conditional formatting tool in MS Excel spreadsheets. The comments in the last column can be done by using the IF statement function. You can sort the scores so they are arranged from highest to lowest or vice versa.

Another important aspect to monitoring a learner's progress can be done as follows:

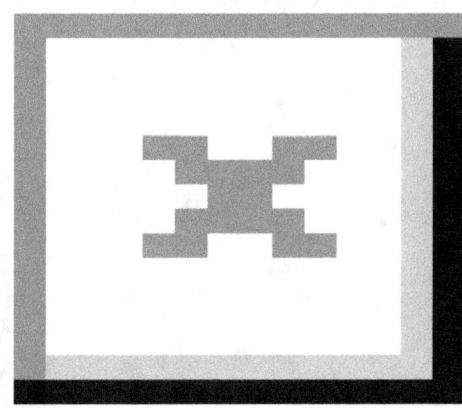

Using MS Excel charts/graphs to compare learners' achievement

In the image above, you can see a comparison between numbers of learners expected grades and the actual grades that learners achieved. This type of data analysis would be useful for the school's administration as well. Teachers could use this type of data to tweak their teaching approaches to achieve more desirable outcomes for their learners next time.

Task 4

Create a spreadsheet workbook to keep learner scores for tasks and tests. Create a separate workbook for each class. Use formulas to work out learner average scores. Apply conditional formatting to them so that you can monitor learners who fall below a certain level.

Scenario 5

Mr Scolian, as the head of department, needs to collect grades from the teachers in his department each term. He has a spreadsheet workbook with all the learners details but asks each teacher to collaborate with him by adding their classes grades into his spreadsheet themselves. This an example of how several teachers can collaborate on MS Excel workbooks in real time.

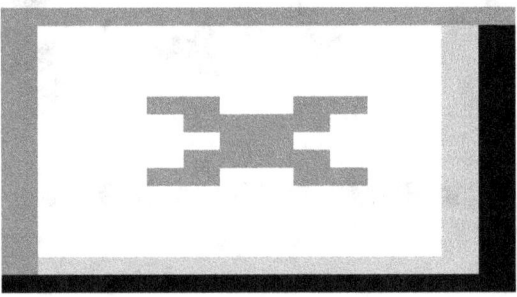

Using charts and graphs to track and share learners' progress

Discussion

Teachers can upload learners grades to collaborate with their colleagues on bigger projects. This feature of 'Share' found at the top right-hand corner of a MS Excel workbook enables you to add the email addresses of colleagues that you wish to share a workbook with. As they enter their data, it will change in real time on your original spreadsheet workbook. Their entries will appear in a different colour and by clicking on one of their cells, you can identify who added the data because their email address will show up.

The benefits of sharing data in real time are many. It saves time and once you've started collaborating in this way, it becomes easier to do. In addition, you can collaborate with others, not just your colleagues.

Task 5

- Create a spreadsheet workbook to collect data from your new students for their extra-curricular activities choices, such as, days they are available, choice of club they wish to join, etc. Use the 'Share' feature by emailing each learner and asking them to complete the section next to their name.
- Describe the problems you experienced doing this exercise.
- Suggest different ways you could use this feature of MS Excel.

Practical Task for Teachers

Imagine your school cannot open due to COVID and you have to still reach out to your learners by sending work to them via Google Classroom. All your learners have access to MS Word, MS Excel and MS PowerPoint and to the internet.

- Describe how you would arrange the first two week's set of lessons for your classes.
- Use MS PowerPoint to set out the lesson for each class showing clearly, the Starter, Main Teaching, Learner Tasks and Plenary.
- Upload the worksheets made in MS Word for learners to carry out their tasks.
- Create a MS Excel spreadsheet to record and keep track of each learner's grades for each assessment or task.
- Use a collaborative spreadsheet workbook to collect feedback from your learners about how well distance learning has worked for them and for suggestions on how they could improve their own outcomes next time and for any suggestions about how they feel you could improve your delivery next time.
- Evaluate your own performance by answering these questions:

- What have I learnt from the learner feedback?
- Is there anything I would do differently next time to improve learner outcomes?
- What would I recommend to my colleagues regarding the benefits of distance teaching and learning based on my two week experience?

Conclusion

1

This book is written in an unconventional way. It is not meant to be read from the start to the end as you would read most publications. It is meant to be experienced and by this I mean, you should read a section, try it out for yourself with your students and colleagues before proceeding to the next section. You will need to use one or more digital gadgets, such as, a SMART phone, tablet, phablet, laptop or desktop computer. You should also have an email address to sign up to most of the APPs and online platforms.

It is advisable to share with your colleagues where possible and in the spirit of today's online share culture. Try to combine the uses of different digital technologies for good effect, such as, creating a collaborative space in, say, TRELLO and then sharing in on a suitable social media site. The spirit of today's working, learning and teaching ethos is also one of transparency, as each entry and deletion is traceable, thereby transparent. This has an unintended benefit of driving up standards and improving outcomes for a group of students, a school or learning institution as a whole and much more.

Each chapter covers some of the more unusual ways that digital technologies can be used as part of a continuing professional development strategy in any educational setting. It's more the impact of their uses that should be a more central focus. As we've experienced a pandemic that had the potential of stopping education as we've known it in the past few centuries, we should be prepared to explore different ways to 'reach and teach'. The current generation are digital natives and as educators we ought to keep up to speed with the latest digital technologies and how they could increase student engagement and enhance the whole teaching and learning experience. It is easy for the technology to become a distraction from the main learning purpose, so this book is an endeavour to introduce you to a host of digital technologies and how to make them a seamless part of your teaching and learning experiences henceforth.

The best way to enjoy this book is to tackle each section as a learning event. Be prepared to sign up to various APPs, online platforms and try out the exercises with your students. Share your lesson designs and ask for feedback from your colleagues and students in an attempt to learn what works best for your students and yourself. You will probably find some sections more challenging than others but if you persist and reach out to your colleagues, you will be amazed at how much everyone will benefit.

About the author

2

A 21st Century E-Learning Guide for Teachers, Trainers and Instructors

Denise Taylor

Denise Taylor is an internationally experienced educator, having been a teacher, head of department, head of school, examiner, moderator and principal examiner for various examination boards, chair of school governors, lecturer at a teacher training college and author of educational textbooks for international audiences. She is qualified and holds a BA, B.Ed and MA(ODE) and has also been involved in educational endeavours in South Africa, UK, China, Ghana, Kazakhstan and Saudi Arabia. She is also a wife and proud mother of four children.

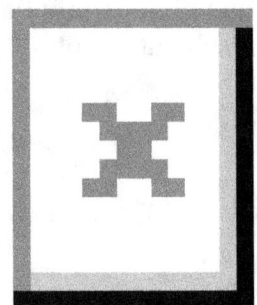